I0493669

Table of Contents

PREFACE

On March 28, 2014 the Obama Administration released a key element called for in the President's Climate Action Plan: a Strategy to Reduce Methane Emissions. The strategy summarizes the sources of methane emissions, commits to new steps to cut emissions of this potent greenhouse gas, and outlines the Administration's efforts to improve the measurement of these emissions. The strategy builds on progress to date and takes steps to further cut methane emissions from several sectors, including the oil and natural gas sector.

This technical white paper is one of those steps. The paper, along with four others, focuses on potentially significant sources of methane and volatile organic compounds (VOCs) in the oil and gas sector, covering emissions and mitigation techniques for both pollutants. The Agency is seeking input from independent experts, along with data and technical information from the public. The EPA will use these technical documents to solidify our understanding of these potentially significant sources, which will allow us to fully evaluate the range of options for cost-effectively cutting VOC and methane waste and emissions.

1.0   INTRODUCTION

The oil and natural gas exploration and production industry in the U.S. is highly dynamic and growing rapidly. Consequently, the number of wells in service and the potential for greater air emissions from oil and natural gas sources is also growing. There were an estimated 504,000 producing gas wells in the U.S. in 2011 (U.S. EIA, 2012a), and an estimated 536,000 producing oil wells in the U.S. in 2011 (U.S. EIA, 2012b). It is anticipated that the number of gas and oil wells will continue to increase substantially in the future because of the continued and expanding use of horizontal drilling combined with hydraulic fracturing (referred to here as simply hydraulic fracturing).

Due to the growth of this sector and the potential for increased air emissions, it is important that the U.S. Environmental Protection Agency (EPA) obtain a clear and accurate understanding of emerging data on air emissions and available mitigation techniques. This paper presents the Agency's understanding of emissions and available emissions mitigation techniques from a potentially significant source of emissions in the oil and natural gas sector.

In new gas wells, there is generally sufficient reservoir pressure to facilitate the flow of water and hydrocarbon liquids to the surface along with produced gas. In mature gas wells, the accumulation of liquids in the well can occur when the bottom well pressure approaches reservoir shut-in pressure. This accumulation of liquids can impede and sometimes halt gas production. When the accumulation of liquid results in the slowing or cessation of gas production (i.e., liquids loading), removal of fluids (i.e., liquids unloading) is required in order to maintain production. Emissions to the atmosphere during liquids unloading events are a potentially significant source of VOC and methane emissions.

Most gas wells will have liquid loading occur at some point during the productive life of the well. When this occurs, common courses of action to improve gas flow include (U.S. EPA, 2011):

- Shutting in the well to allow bottom hole pressure to increase, then venting the well to the atmosphere (well blowdown, or "blowing down the well"),

- Swabbing the well to remove accumulated fluids,
- Installing a plunger lift,
- Installing velocity tubing, and
- Installing an artificial lift system.

Blowing down the well involves the intentional manual venting of the well to the atmosphere to improve gas flow, whereas the use of a plunger lift system uses the well's own energy (gas/pressure) to lift liquids from the tubing by pushing the liquids to the surface by the movement of a free-traveling plunger ascending from the bottom of the well to the surface. The plunger essentially acts as a piston between liquid and gas. Use of a plunger lift often minimizes and sometimes eliminates the need for blowing down the well.

Because of the potential for substantial VOC and methane emissions occurring during liquids unloading at natural gas wells, there are an increasing number of studies on emissions from natural gas well liquids unloading events. These studies of liquids unloading practices attempt to quantify emissions on a well specific, regional and national level and often take into account the use of available mitigation techniques, such as plunger lifts. This document provides a summary of the EPA's understanding of VOC and methane emissions from natural gas production liquids unloading events, available liquids unloading and emission mitigation techniques, the relative magnitude of emissions associated with the respective techniques and the efficacy and prevalence of those techniques in the field. Section 2 of this document provides our understanding of emissions from liquids unloading events, and Section 3 provides our understanding of available liquids unloading and emissions mitigation techniques. Section 4 summarizes the EPA's understanding based on the information presented in Sections 2 and 3, and Section 5 presents a list of charge questions for reviewers to assist us with obtaining a more comprehensive understanding of liquids unloading VOC and methane emissions and emission mitigation techniques for the liquids unloading process.

2.0 OIL AND NATURAL GAS SECTOR LIQUIDS UNLOADING AVAILABLE EMISSIONS DATA AND EMISSIONS ESTIMATES

Given the potential for significant emissions from liquids unloading, there have been several information collection efforts and studies conducted to estimate emissions and available emission mitigation techniques. Some of these studies are listed in Table 2-1, along with an indication of the type of information contained in the study (i.e., activity level, emissions data, mitigation techniques).

Table 2-1. Summary of Major Sources of Liquids Unloading Information

Name	Affiliation	Year of Report	Activity Data	Emissions Data	Mitigation Techniques
Greenhouse Gas Reporting Program (U.S. EPA, 2013)	U.S. Environmental Protection Agency	2013	Sub-basin	X	X
Inventory of Greenhouse Gas Emissions and Sinks: 1990-2012 (2014 GHG Inventory) (U.S. EPA, 2014)	U.S. Environmental Protection Agency	2013	Regional	X	X
Characterizing Pivotal Sources of Methane Emissions from Natural Gas Production: Summary and Analysis of API and ANGA Survey Responses (API and ANGA, 2012)	American Petroleum Institute (API)/America's Natural Gas Alliance (ANGA)	2012	Regional	X	X
Measurements of Methane Emissions at Natural Gas Production Sites in the United States (Allen et al., 2013)	Multiple Affiliations, Academic and Private	2013	9 Liquids Unloading Events	X	X
Economic Analysis of Methane Emission Reduction Opportunities in the U.S. Onshore Oil and Natural Gas Industries (ICF International, 2014)	ICF International (Prepared for the Environmental Defense Fund)	2014	Regional	X	X

A more-detailed description of the data sources listed in Table 2-1 is presented in the following sections, including how the data may be used to estimate national VOC and methane emissions from liquids unloading events.

In October 2013, the EPA released 2012 greenhouse gas (GHG) data for Petroleum and Natural Gas Systems[1] collected under the Greenhouse Gas Reporting Program (GHGRP). The GHGRP, which was required by Congress in the FY2008 Consolidated Appropriations Act, requires facilities to report data from large emission sources across a range of industry sectors, as well as suppliers of certain GHGs and products that would emit GHGs if released or combusted.

When reviewing this data and comparing it to other datasets or published literature, it is important to understand the GHGRP reporting requirements and the impacts of these requirements on the reported data. The GHGRP covers a subset of national emissions from Petroleum and Natural Gas Systems; a facility[2] in the Petroleum and Natural Gas Systems source category is required to submit annual reports if total emissions are 25,000 metric tons carbon dioxide equivalent (CO_2e) or more. Facilities use uniform methods prescribed by the EPA to calculate GHG emissions, such as direct measurement, engineering calculations, or emission factors derived from direct measurement. In some cases, facilities have a choice of calculation methods for an emission source.

The liquids unloading source emissions reported under the GHGRP include emissions from facilities that have wells that are venting, including those wells that vent during plunger lift operation. Liquids unloading techniques that do not involve venting are not reported. The total reported methane emissions in 2012 for liquids unloading were approximately 276,378 metric tons (MT). Facilities were given the option among three methods for calculating emissions from liquids unloading. The first calculation method involved using a representative well sample to calculate emissions for both wells with and without plunger lifts. The second and third

---

[1] The implementing regulations of the Petroleum and Natural Gas Systems source category of the GHGRP are located at 40 CFR Part 98 Subpart W.

[2] In general, a "facility" for purposes of the GHGRP means all co-located emission sources that are commonly owned or operated. However, the GHGRP has developed a specialized facility definition for onshore production. For onshore production, the "facility" includes all emissions associated with wells owned or operated by a single company in a specific hydrocarbon producing basin (as defined by the geologic provinces published by the American Association of Petroleum Geologists).

calculation methods provided engineering equations for wells with plunger lifts and without plunger lifts.

Of the 251 facilities that reported emissions for well venting for liquids unloading, 120 facilities reported using Best Available Monitoring Methods (BAMM) for unique or unusual circumstances. Where a facility used BAMM, it was required to follow emission calculations specified by the EPA, but was allowed to use alternative methods for determining inputs to calculate emissions. These inputs are the values used by facilities to calculate equation outputs or results. Table 2-2 shows the activity count and reported emissions for the different calculation methods.

Table 2-2. Greenhouse Gas Reporting Program 2012 Reported Emissions from Liquids Unloading (U.S. EPA, 2013)

Calculation Method	Number of Facilities Reporting[a]	Number of Wells Venting During Liquids Unloading	Number of Wells Venting that are Equipped With Plunger Lifts	Reported CH_4 Emissions (MT)[b]
Method 1: Direct Measurement of Representative Well Sample	42	10,024	7,149	112,496
Method 2: Engineering Calculation for Wells without Plunger Lifts	188	23,536	0	71,593
Method 3: Engineering Calculation for Wells with Plunger Lifts	132	25,103	25,103	92,289
Total	251	58,663	32,252	276,378

[a] Total number of facilities is smaller than the sum of facilities from each method because some facilities reported under both Method 2 and Method 3.
[b] The reported CH_4 MT CO_2e emissions were converted to CH_4 emissions in MT by dividing by a global warming potential (GWP) of methane (21).

2.2   API/ANGA 2012 Survey Data (API and ANGA, 2012)

The API/ANGA 2012 Survey Data includes a dataset from over 20 companies covering over 90,000 gas wells, including approximately 59,000 wells that conducted liquids unloading operations. This study sample population includes representation from most of the geographic regions of the country as well as most of the geologic formations currently developed by the industry. The study provides estimated methane emissions from liquids unloading for 5,327 wells that were calculated based on well characteristics such as well bore volume, well pressure, venting time, and gas production rate and using 40 CFR part 98, subpart W engineering equations. These emissions estimates and the activity data used to calculate the estimates are presented in Table 2-3.

Table 2-3. API/ANGA Study Liquids Unloading Emissions Estimates
(API and ANGA, 2012; pg. 14)

Mid-Level Survey Data	
Total number of wells with plunger lift (42,681 in sample)	11,518
Total number of wells without plunger lift (42,681 in sample)	31,163
Number of plunger equipped wells that vent (42,681 in sample)	2,426 (21.1%)
Number of non-plunger equipped wells that vent (i.e., wells performing blowdowns)(42,681 in sample)	2,901 (9.3%)
Total annual volume gas vented for venting wells	1,719,843,596 standard cubic feet (scf) gas/year
Calculated volume vented gas per venting well	322,854 scfy gas/well
Calculated methane volume vented per venting well	254,409 scfy CH₄/well
Calculated National Well Data	
Calculated national number of wells with plunger lift that vent for unloading	36,806
Calculated national number of wells without plunger lift that vent for unloading (i.e., wells performing blowdowns)	28,863
National Emission Calculations	
Total gas venting for liquids unloading volume (scaled for national wells)	21,201,410,618 scf gas/yr
Total methane venting for liquids unloading (scaled for national wells)	16,706,711,567 scf CH4/yr
Total liquid unloading vented methane (scaled for national wells)	319,664 MT CH4/yr

The authors of the study made the following conclusions:

- The 2012 GHG Inventory emissions estimates for liquids unloading were overestimated by orders of magnitude. The API/ANGA Survey data indicated a lower percentage of gas wells that vent for liquids unloading and a shorter vent duration.

- The emissions from liquids unloading are not specific to only conventional wells, but can be for any well depending on several technical and geological aspects of the well.

- Although most wells do not require liquids unloading until later in the well's productive lifetime, the timeframe for initiating liquids unloading operations varies significantly and can be early in the well's productive life span.

- Most of the emissions from liquids unloading operations are produced by less than 10% of the venting well population.

The study does not discuss the characteristics that cause certain wells to have significantly higher emissions than other venting wells. The study showed that the majority of emissions came from a small percentage of venting wells, and both conventional and hydraulically fractured wells can vent during liquids unloading. Additionally, while a large percentage of wells equipped with plunger lifts do not vent during liquids unloading events, many wells with plunger lifts produce emissions during liquids unloading events. This suggests that plunger lifts are capable of unloading liquids from a well without venting, but in many cases they are operated in a manner that results in venting. It is not clear to the EPA what the conditions are that cause these wells with plunger lifts to be operated in a manner that results in significant venting during liquids unloading.

2.3   Inventory of U.S. Greenhouse Gas Emissions and Sinks: 1990-2012 (U.S. EPA, 2014)

The EPA leads the development of the annual Inventory of U.S. Greenhouse Gas Emissions and Sinks (GHG Inventory). This report tracks total U.S. GHG emissions and removals by source and by economic sector over a time series, beginning with 1990. The U.S. submits the GHG Inventory to the United Nations Framework Convention on Climate Change (UNFCCC) as an annual reporting requirement. The GHG Inventory includes estimates of methane and carbon dioxide for natural gas systems (production through distribution) and petroleum systems (production through refining). The 2014 GHG Inventory data (published in 2014; containing emissions data for 1990-2012) was evaluated for information on liquids unloading emissions.

The 2014 GHG Inventory applied calculated National Energy Modeling System (NEMS) (U.S. EPA, 2014) region- and unloading technology-specific emission factors to the percentage of wells requiring liquids unloading by using the percentages of wells venting for liquids unloading with plunger lifts, and wells without plunger lifts in each region based on API/ANGA Survey data (*see* Section 2.1.1.3 for a discussion on this data).

The 2014 GHG Inventory activity data (number of wells), emissions factors (standard cubic feet per year [scfy]/well) and the calculated emissions for liquids unloading are presented by NEMS region in Table 2-4.

Table 2-4. Data and Calculated CH$_4$ Emissions [MT] for the Natural Gas Production Sector by NEMS Region (U.S. EPA, 2014, ANNEX 3 Methodological Descriptions for Additional Source or Sink Categories)

NEMS Region	Activity	Activity Data[a,b] (number of wells)	Emission Factor (scfy)/well[b]	Calculated Emissions (MT)
North East	Liquids Unloading (with plunger lifts)	6,924	268,185	35,764
	Liquids Unloading (without plunger lifts; blowdowns)	17,906	141,646	48,849
Midcontinent	Liquids Unloading (with plunger lifts)	2,516	1,140,052	55,245
	Liquids Unloading (without plunger lifts; blowdowns)	4,469	190,179	16,369
Rocky Mountain	Liquids Unloading (with plunger lifts)	10,741	119,523	24,726
	Liquids Unloading (without plunger lifts; blowdowns)	1,267	1,998,082	48,758

NEMS Region	Activity	Activity Data[a,b] (number of wells)	Emission Factor (scfy)/well[b]	Calculated Emissions (MT)
South West	Liquids Unloading (with plunger lifts)	1,379	2,856	76
	Liquids Unloading (without plunger lifts; blowdowns)	8,078	77,899	12,120
West Coast	Liquids Unloading (with plunger lifts)	159	317,292	972
	Liquids Unloading (without plunger lifts; blowdowns)	142	279,351	764
Gulf Coast	Liquids Unloading (with plunger lifts)	1,784	61,758	2,122
	Liquids Unloading (without plunger lifts; blowdowns)	5,445	265,120	27,803
Total		60,810		273,568

[a]DI Desktop, 2014.
[b]API/ANGA 2012 Survey Data, Characterizing Pivotal Sources of Methane Emissions from Natural Gas Production – Summary and Analysis of API and ANGA Survey Responses (API and ANGA, 2012).

The 2014 GHG Inventory data estimates that liquids unloading emissions in 2012 were 14% of overall methane emissions from the natural gas production segment.

2.4 Measurements of Methane Emissions at Natural Gas Production Sites in the United States (Allen et al., 2013)

A study completed by multiple academic institutions and consulting firms was conducted to gather methane emissions data at onshore natural gas sites in the U.S. and compare those emission estimates to the 2011 estimates reported in the EPA's 2013 GHG Inventory. The sources or operations tested included liquids unloading. Under this study, sampling was performed for liquids unloading in which an operator manually bypassed the well's separator. These manual unloading events could be scheduled, which allowed time to install measurement equipment.

Analysis included nine well unloading events, ranging from 15 minutes to two hours, including both continuous flow and intermittent flow events. Some of the wells sampled only unloaded liquids once over the current life of the well, where others were unloaded monthly. The average emissions per unloading event were 1.1 MT of methane (95% confidence limits of 0.32-2.0 MT). The study reports that the average emissions per well per year (based on the emissions per event for each well multiplied by the frequency of the events per year reported by the well operator) was 5.8 MT. The study acknowledges that the sampled population characteristics reflected a wide range of emission rates and that when emissions are averaged per event, emissions from four of the nine events contribute more than 95% of the total emissions. This result is consistent with the API/ANGA 2012 Survey Data and 2012 data reported to the GHGRP; all suggest that certain wells produce significantly more emissions during liquids unloading events than others. The study also suggests that the length of the liquids unloading event and the number of events are crucial factors in a well's annual emissions from liquids unloading.

The authors report that their study supports their belief that the application of the API/ANGA 2012 Survey data method used when calculating the 2013 GHG Inventory overestimates GHG emissions. Although the authors believe that their study provides valuable information, they caution that the sampling in their study was insufficient to characterize emissions from liquids unloading for all well sites in all basins and recommend that additional measurement of unloading events be conducted in order to improve national emissions estimates. Because characteristics of the unloading events sampled in the study were highly variable, and because the number of events sampled was small, the authors caution the use of the data to extrapolate to larger populations.

2.5 Economic Analysis of Methane Emission Reduction Opportunities in the U.S. Onshore Oil and Natural Gas Industries (ICF International, 2014)

The Environmental Defense Fund (EDF) commissioned ICF International (ICF) to conduct an economic analysis of methane emission reduction opportunities from the oil and natural gas industry to identify the most cost-effective approach to reduce methane emissions from the industry. The study projects the estimated growth of methane emissions through 2018 and focuses its analysis on 22 methane emission sources in the oil and natural gas industry (referred to as the targeted emission sources). These targeted emission sources represent 80% of their projected 2018 methane emissions from onshore oil and gas industry sources. Liquids unloading is one of the 22 emission sources that is included in the study.

The study relied on the EPA's 2013 GHG Inventory for methane emissions data for the oil and natural gas sector. This emissions data was revised to include updated information from the GHGRP (EPA) and the *Measurements of Methane Emissions at Natural Gas Production Sites in the United States* study (Allen et al., 2013). The revised 2011 baseline methane emissions estimate was used as the basis for projecting onshore methane emissions to 2018. The projected emissions are not discussed further here, because projected emissions are not a topic covered by this white paper.

The study used the GHGRP data for 2011 and 2012 to develop new activity and emission factors for wells with liquids unloading. It was assumed that the respondents represented 85% of the industry, therefore, the EPA's 2013 GHG Inventory estimate of the number of venting wells with plunger lifts was increased to 44,286 from 37,643, and the estimate of the number of venting wells without plunger lifts was increased to 31,113 from 26,451.[3] The emission factors were updated by dividing the total emissions for each venting well type (those equipped with plunger lifts and those that were not equipped with a plunger lift) by the total number of reporting wells. The calculated emission factors were 277,000 scf/venting well for wells with plunger lifts and 163,000 scf/venting well for wells without plunger lifts. Using these updated emission factors, ICF estimated a net increase of methane emissions from liquids unloading (as compared to the EPA's 2013 GHG Inventory) by approximately 30% to 17 billion cubic feet (Bcf)(approximately 321,012 MT). This represented the study's baseline methane emissions for 2011 for liquids unloading.

---

[3] The EPA is unaware of how the study authors determined the GHGRP data represented 85% of the industry.

Further information included in this study on the use of a plunger lift as a mitigation or emission reduction option, methane control costs, and their estimates for the potential for VOC emissions co-control benefits from the use of a plunger lift are presented in Section 3.1 of this document.

3.0 AVAILABLE LIQUIDS UNLOADING EMISSIONS MITIGATION TECHNIQUES

As noted previously, many natural gas wells have sufficient reservoir pressure to flow formation fluids (water and hydrocarbon liquids) to the surface along with the produced gas. As the bottom well pressure approaches reservoir shut-in pressure, gas flow slows and liquids accumulate at the bottom of the tubing. A common approach to temporarily restoring flow is to vent the well to the atmosphere (well "blowdown") which removes liquids but also produces emissions.

Several techniques are available that could produce less (compared to blowdown) or no emissions from liquids unloading. The following section describes techniques that remove liquids from the well by other means than a blowdown and in the process can reduce the amount of vented gas and, thus, reduce the VOC and methane emissions. These technologies can reduce the need for liquids unloading operations or result in the capture of gas from liquids unloading operations.

3.1 Liquid Removal Technologies

Numerous liquid removal technologies have been evaluated for their emission levels and their potential for eliminating or minimizing emissions from liquids unloading. The Natural Gas STAR program reports the potential for significant emissions reductions and economic benefits from implementing one or more lift options to remove this liquid instead of blowing down the well to the atmosphere (U.S. EPA, 2006b and 2011).

As noted in Section 1 of this document, the Natural Gas STAR program reports that when liquids loading occurs during the productive life of the well, one or more of the following actions are generally taken (U.S. EPA, 2011):

- Shutting in the well to allow the bottom hole pressure to increase, and then venting the well to the atmosphere (well blowdown),
- Swabbing the well to remove accumulated fluids,

- Installing velocity tubing,

- Installing a plunger lift system, and

- Installing an artificial lift system.

In the sections below, the technologies have been divided into "primary" and "remedial" technologies. It is the EPA's understanding that the "primary" technologies are used as more permanent solutions to liquids loading problems, while the "remedial" technologies may mitigate the problem but do not provide a long term permanent solution. These technologies are summarized in Table 3-1.

Table 3-1. Liquid Removal Techniques for Liquid Unloading of Natural Gas Wells

Mitigation Techniques	Description	Applicability	Costs	Efficacy and Prevalence
Primary Techniques				
Plunger Lift Systems	Plunger lifts use the well's own energy (gas/pressure) to drive a piston or plunger that travels the length of the tubing in order to push accumulated liquids in the tubing to the surface (U.S. EPA, 2006b).	Plunger systems have been known to reduce emissions from venting and increase well production. Specific criteria regarding well pressure and liquid to gas ratio can affect applicability. Candidate wells for plunger lift systems generally do not have adequate downhole pressure for the well to flow freely into a gas gathering system (U.S. EPA, 2006b).	The following information is from the EPA's Natural Gas STAR Program technical documents, however, additional cost data may be available such as from equipment or service providers (U.S. EPA, 2006b and 2011): • Capital, installation and startup cost estimates: $1,900-$7,800 (Note: Commenters on the ICF study cited a cost of $15,000. The study escalated the cost to $20,000 (ICF International, 2014)) • Smart automation system: $4,700/well - $18,000/well depending on the complexity of the system. • Additional startup costs (e.g., well depth survey, miscellaneous well clean out operations): $700-$2,600.	API/ANGA Survey data show plunger lifts can result in zero emissions or significant emissions depending on how they are operated. The EPA has learned plunger lift systems rely on manual, onsite adjustments. When a lift becomes overloaded, the well must be manually vented to the atmosphere to restart the plunger. Optimized plunger lift systems (e.g., with smart well automation) can decrease the amount of gas vented by up to 90+% and reduce the need for venting due to overloading (U.S. EPA, 2006b).

16

Mitigation Techniques	Description	Applicability	Costs	Efficacy and Prevalence
			• Annual operating and maintenance costs (e.g., inspection and replacement of lubricator and plunger): $700-$1,300 • Annual cost savings from avoided emissions from use of an automated system: $2,400-$10,241.	
Artificial lifts (e.g., rod pumps, beam lift pumps, pumpjacks and downhole separator pumps)	Artificial lifts require an external power source to operate a pump that removes the liquid buildup from the well tubing (U.S. EPA, 2011).	The devices are typically used during the eventual decline in the gas reservoir shut-in pressure, when there is inadequate pressure to use a plunger lift. At this point, the only means of liquids unloading to keep gas flowing is downhole pump technology (U.S. EPA, 2011).	The following information is from the EPA's Natural Gas STAR Program technical documents, however, additional cost data may be available such as from equipment or service providers (U.S. EPA, 2011): • Capital and installation costs (includes location preparation, well clean out, artificial lift equipment and pumping unit): $41,000-$62,000/well • Average cost of pumping unit: $17,000-$27,000.	Artificial lifts can be operated in a manner that produces no emissions (U.S. EPA, 2011). The EPA does not have information on the prevalence of this technology in the field.

Technique	Description	Applicability / Operation	Cost	Comments
Velocity tubing	Velocity tubing is smaller diameter production tubing and reduces the cross-sectional area of flow, increasing the flow velocity and achieving liquid removal without blowing emissions to the atmosphere. Generally, a gas flow velocity of 1,000 feet per minute (fpm) is necessary to remove wellbore liquids (U.S. EPA, 2011).	• Velocity tubing strings are appropriate for low volume natural gas wells upon initial completion or near the end of their productive lives with relatively small liquid production and higher reservoir pressure. Candidate wells include marginal gas wells producing less than 60 Mcfd (U.S. EPA, 2011). • Coil tubing can also be used in wells with lower velocity gas production (U.S. EPA, 2011).	The following information is from the EPA's Natural Gas STAR Program technical documents, however, additional cost data may be available such as from equipment or service providers (U.S. EPA, 2011): • Installation requires a well workover rig to remove the existing production tubing and place the smaller diameter tubing string in the well. • Capital and installation costs provided from industry include the following: $7,000-$64,000/well	Considered to be a "no emissions" solution. Low maintenance, effective for low volumes lifted. Often deployed in combination with foaming agents. Seamed coiled tubing may provide better lift due to elimination of turbulence in the flow stream (U.S. EPA, 2011). The EPA does not have information on the prevalence of this technology in the field.
Foaming agents	A foaming agent (soap, surfactants) is injected in the casing/tubing annulus by a chemical pump on a timer basis. The gas bubbling through the soap-water solution creates gas-water foam which is more easily lifted to the surface for water removal (U.S. EPA, 2011).	A means of power will be required to run the surface injection pump. The soap supply will also need to be monitored. If the well is still unable to unload fluid, additional, smaller tubing may be needed to help lift the fluids. Foaming agents work best if the fluid in the well is at least 50% water. Surfactants are not effective for natural gas liquids or liquid hydrocarbons. Foaming agents and velocity tubing may be more effective when used in combination (U.S. EPA, 2011).	The following information is from the EPA's Natural Gas STAR Program technical documents, however, additional cost data may be available such as from equipment or service providers (U.S. EPA, 2011): Foaming agents are low cost. No equipment is required in shallow wells. In deep wells, a surfactant	Considered to be a "no emissions" solution. Low volume method applied early in production decline when bottom hole pressure still generates sufficient velocity to lift liquid droplets (U.S. EPA, 2011). The EPA does not have information on the prevalence of this technology in the field.

injection system requires the installation of surface equipment and regular monitoring. Pump can be powered by solar or AC power or actuated by movement of another piece of equipment.

- Capital and startup costs to install soap launchers: $500-$3,880
- Capital and startup costs to install soap launchers and velocity tubing: $7,500-$67,880
- Monthly cost of foaming agent: $500/well or $6,000/yr

3.1.1   Primary Techniques

Plunger Lifts

Based on our assessment of the data, a plunger lift system for liquids unloading is capable of performing liquids unloading with little or no emissions. The level of emissions depends on how the plunger lift system is operated, specifically, whether gas is directed to the sales line or vented to the atmosphere. There may be potential for improved production and emissions reduction when paired with a smart well automation that optimizes production and reduces product losses to the atmosphere. A schematic diagram of a plunger lift is shown in Figure 3-1.

Basic installation costs for plunger lifts were estimated as ranging from $1,900 - $7,800 based on information gathered from the EPA's Natural Gas STAR program (*see* Table 3-1). Plunger lift installation costs include installing the piping, valves, controller and power supply on the wellhead and setting the downhole plunger bumper assembly, assuming the well tubing is open and clear. Lower costs (*e.g.*, $1,900) would result where no other activities are required for installation. Higher costs (*e.g.*, $7,800) would be incurred in situations where running a wire-line, which is necessary to check for internal blockages within the tubing, and a test run of the plunger is conducted from top to bottom (a process also known as broaching) to ensure that the plunger will move freely up and down the tubing string (U.S. EPA, 2006b).

Other startup costs in addition to the installation costs can include a well depth survey, swabbing to remove well bore fluids, removing mineral scale and cleaning out perforations, fishing out debris in the well, and other miscellaneous well clean out operations. Additional startup costs were estimated to be $700 - $2,600 (U.S. EPA, 2006b). However, commenters on the ICF study cited startup costs of $15,000. The commenters also stated that well treatments and clean outs are often required before plunger lifts can be installed. The study escalated the cost to $20,000 per well (ICF International, 2014).

Figure 3-1. Example Plunger Lift (U.S. EPA, 2006b)

The activities to install the smart automation plunger lift include installing the controller, power supply, and host system, in addition to the activities required for the plunger itself. The typical cost of automating a plunger lift system is approximately $5,700 - $18,000, depending on the complexity of the well. This cost would be in addition to the startup costs of a plunger-only system (U.S. EPA, 2011). Installing telemetry units can help to optimize production; however, automated controllers are not necessarily required for reducing emissions.

Natural Gas STAR Partners have reported methane emissions reductions and economic benefits from implementing plunger lifts as compared to conducting blowdowns, especially those equipped with smart automation systems. The reported economic benefits from natural gas savings and improved well production range from $2,400 - $4,389 per well per year[4] (U.S. EPA, 2011). The EPA is not aware of any adverse secondary environmental impacts that would result from the installation and operation of plunger lifts in a liquid producing natural gas well, and the use of a smart automated plunger lift system has the potential to optimize production and minimize emissions over the use of a non-automated plunger lift system.

The ICF study (ICF International, 2014) calculated emission control cost curves ($/Mcf of methane reduced) using their 2018 projected methane emission estimates. The primary sources used for projecting onshore methane emissions for liquids unloading for 2018 included natural gas forecast information from the U.S. EIA's Annual Energy Outlook (AEO) and 2014 Early Release (*Lower 48 Natural Gas Production and Supply Prices by Supply Region*) and API's Quarterly Well Completions Report. The EIA information was used to project methane emissions by using regional gas production information projected in EIA's 2014 AEO for 2018. The API's report was used to update well counts by EIA AEO regions whereby a ratio of the number of wells in 2018 to wells in 2011 was used to drive the activity for most of the emission sources involved in gas production. The study assumed the application of a plunger lift (assuming 95% control of methane emissions) on 30% of the estimated venting wells without plunger lifts. ICF estimated a methane reduction of 1.6 Bcf (or approximately 30,212 MT) at a cost of $0.74/Mcf methane reduced with the application of a plunger lift on these uncontrolled wells. ICF also estimated that VOC emissions would be reduced by 9.3 kilotons (or approximately 9,300 MT) at a cost of $125/ton. According to the report, liquids unloading can increase production by anywhere from 3 to 300 thousand cubic feet per day (Mcf/day) and, without taking credit for the productivity increase, the report estimates that the cost-effectiveness breakeven point is about 1,200 Mcf/yr of venting (estimated as a reduction cost of $0.05/Mcf reduced). Their analysis assumed capital costs of $20,000 and annual operating costs of $2,400.

Artificial Lift Systems

Artificial lift systems (e.g., rod pumps and pumping units) require an external power source to operate, such as electric motors or natural gas fueled engines. However, these systems can be

---

[4] Assumes a gas price of $3 per Mcf.

installed and effectively remove liquids from the well even after the well pressure has declined to the point where a plunger lift system can no longer be operated, thus they are capable of prolonging the life of a well. They typically require the use of a well workover rig to install a downhole rod pump, rods, and tubing in the well.

Based on results reported by Natural Gas STAR Partners, the cost of implementing artificial lift systems range from $41,000 - $62,000. The reported economic benefits from natural gas savings range from $2,919 - $6,120 per well per year[5] (U.S. EPA, 2011).

Secondary environmental impacts occur due to the emissions from the natural-gas fueled engine used to power the lift system, however, these impacts can be reduced by using an electric motor instead.

3.1.2   Remedial Techniques

Velocity Tubing

As was described previously, liquids build up in the well tubing as well pressure declines and the gas flow velocity is not sufficient to push the liquids out of the well tubing. Velocity tubing (smaller diameter production tubing) decreases the cross-sectional area of the conduit through which the gas flows and thus increases the velocity of the flow. The Natural Gas STAR Program uses 1,000 fpm as a general rule of thumb for the velocity necessary to remove liquids from the well (Note: This is a rule of thumb and the actual required velocity will differ based on well characteristics). When velocity tubing is installed, it must be a small enough diameter to increase the gas flow velocity to 1,000 fpm or to the necessary velocity to remove the liquids from the particular well. A well workover rig is required to remove the existing production tubing and replace it with the velocity tubing. The EPA experience through the Natural Gas STAR Program suggests the wells that are the best candidates for this technology are marginal wells that produce less than 60 Mcfd. However, as well pressure continues to decline as the well ages, the installed velocity tubing may no longer be sufficient to increase the gas flow velocity to the level necessary to remove liquids from the well. At this point, velocity tubing of a smaller diameter or other liquids

---

[5] Estimate does not include value of improved well production. Assumes a gas price of $3 per Mcf.

removal technologies may be required to remove liquids from the well tubing.

Based on results reported by Natural Gas STAR Partners, the cost of implementing velocity tubing ranges from $7,000 - $64,000. The reported economic benefits from natural gas savings and improved well production range from $27,855 - $82,830 per well per year[6] (U.S. EPA, 2011). The EPA is not aware of any adverse secondary environmental impacts that would result from the installation and operation of velocity tubing.

Foaming Agents

Foaming agents can help to remove liquids from wells that are accumulating liquids at low rates. The foam produced from surfactants can reduce the density of the liquid in the well tubing and can also reduce the surface tension of the fluid column, which reduces the gas flow velocity necessary for pushing the liquid out of the well tubing. This technology can be used in conjunction with velocity tubing. However, foaming agents work best when the majority of the liquid built up in the well tubing is water, because they are not effective on natural gas liquids or liquid hydrocarbons (U.S. EPA, 2011).

The foaming agent can be delivered into the well as a soap stick or it can be injected into the casing-tubing annulus or a capillary tubing string. If the well is deep, then an injection system is required that includes foaming agent reservoir, an injection pump, a motor valve with a timer and a power source for the pump (e.g., AC power for electric power or gas for pneumatic pumps) (U.S. EPA, 2011).

Based on results reported by Natural Gas STAR Partners, the costs of foaming agents range from $500 - $9,880. The reported economic benefits from natural gas savings and improved well production range from $1,500 - $28,080 per well per year[7] (U.S. EPA, 2011).

---

[6] Assumes a gas price of $3 per Mcf.
[7] Assumes a gas price of $3 per Mcf.

For deep wells that require an injection system, secondary environmental impacts occur due to the emissions from the power source for the pump. Pneumatic pumps can result in vented gas emissions and electric pumps emissions depending on the source of the electric power.

4.0   SUMMARY

The EPA has used the data sources, analyses and studies discussed in this paper to form the Agency's understanding of VOC and methane emissions from liquids unloading and the emissions mitigation techniques. The following are characteristics the Agency believes are important to understanding this source of VOC and methane emissions:

- A majority of gas wells (conventional and unconventional) must perform liquids unloading at some point during the well's lifetime. As gas wells age and well pressure declines, the need for liquids unloading to enhance well performance becomes more likely.

- The 2014 GHG Inventory estimates the 2012 liquids unloading emissions to be 14% of natural gas production sector emissions.

- The majority of emissions from liquids unloading events come from a small percentage of wells. Some of the characteristics that affect the magnitude of liquids unloading annual emissions from a well are the length of time of each event and the frequency of events.

- A wide range of emission rates from blowdowns have been reported from the limited available well-level data. In the Allen et. al. study, when emissions are averaged per event, emissions from four of the nine events included in the study contribute more than 95% of the total emissions. This result is consistent with the API/ANGA 2012 Survey data and 2012 data reported to the GHGRP; all suggest that certain wells produce more emissions during blowdowns than others. Some suggested causes of this variation are the length of the blowdown and the number of blowdowns per year, which are affected by underlying geologic factors.

- Industry has developed several technologies that effectively remove liquids from wells and can result in fewer emissions than blowdowns. Plunger lifts are the most common of those technologies.

- The emissions reduction efficiency plunger lifts can achieve varies greatly depending on how the system is operated. It is not clear to the EPA what the conditions are that lead to wells with plunger lifts to be vented during plunger lift operation.
- The two liquids unloading techniques that result in vented emissions that the EPA is aware of are plunger lifts when vented to the atmosphere and blowdowns.

5.0   CHARGE QUESTIONS FOR REVIEWERS

1. Please comment on the national estimates of methane emissions and methane emission factors for liquids unloading presented in this paper. Please comment on regional variability and the factors that influence regional differences in VOC and methane emissions from liquids unloading. What factors influence frequency and duration of liquids unloading (e.g., regional geology)?

2. Is there further information available on VOC or methane emissions from the various liquids unloading practices and technologies described in this paper?

3. Please comment on the types of wells that have the highest tendency to develop liquids loading. It is the EPA's understanding that liquids loading becomes more likely as wells age and well pressure declines. Is this only a problem for wells further down their decline curve or can wells develop liquids loading problems relatively quickly under certain situations? Are certain wells (or wells in certain basins) more prone to developing liquids loading problems, such as hydraulically fractured wells versus conventional wells or horizontal wells versus vertical wells?

4. Did this paper capture the full range of feasible liquids unloading technologies and their associated emissions? Please comment on the costs of these technologies. Please comment on the emission reductions achieved by these technologies. How does the well's life cycle affect the applicability of these technologies?

5. Please provide any data or information you are aware of regarding the prevalence of these technologies in the field.

6. In general, please comment on the ability of plunger lift systems to perform liquids unloading

without any air emissions. Are there situations where plunger lifts have to vent to the atmosphere? Are these instances only due to operator error and malfunction or are there operational situations where it is necessary in order for the plunger lift to effectively remove the liquid buildup from the well tubing?

7. Based on anecdotal experience provided by industry and vendors, the blowdown of a well removes about 15% of the liquid, while a plunger lift removes up to 100% (BP, 2006). Please discuss the efficacy of plunger lifts at removing liquids from wells and the conditions that may limit the efficacy.

8. Please comment on the pros and cons of installing a plunger lift system during initial well construction versus later in the well's life. Are there cost savings associated with installing the plunger lift system during initial well construction?

9. Please comment on the pros and cons of installing a "smart" automation system as part of a plunger lift system. Do these technologies, in combination with customized control software, improve performance and reduce emissions?

10. Please comment on the feasibility of the use of artificial lift systems during liquids unloading operations. Please be specific to the types of wells where artificial lift systems are feasible, as well as what situations or well characteristics discourage the use of artificial lift systems. Are there standard criteria that apply?

11. The EPA is aware that in areas where the produced gas has a high H_2S concentration combustion devices/flares are used during liquids unloading operations to control vented emissions as a safety precaution. However, the EPA is not aware of any instances where combustion devices/flares are used during liquids unloading operations to reduce VOC or methane emissions. Please comment on the feasibility of the use of combustion devices/flares during liquids unloading operations. Please be specific to the types of wells where combustion devices/flares are feasible. Are there operational or technical situations where combustion devices/flares could not be used?

12. Given that liquids unloading may only be required intermittently at many wells, is the use of a mobile combustion device/flare feasible and potentially less costly than a permanent combustion device/flare?

13. Given that there are multiple technologies, including plunger lifts, downhole pumps and velocity tubing that are more effective at removing liquids from the well tubing than blowdowns, why do owners and operators of wells choose to perform blowdowns instead of employing one of these technologies? Are there technical reasons other than cost that preclude the use of these technologies at certain wells?

14. Are there ongoing or planned studies that will substantially improve the current understanding of VOC and methane emissions from liquids unloading events and available options for increased product recovery and emissions reductions? The EPA is aware of an additional stage of the Allen et al. study to be completed in partnership with the EDF and other partners that will directly meter the emissions from liquids unloading events. However, the EPA is not aware of any other ongoing or planned studies addressing this source of emissions.

6.0   REFERENCES

Allen, David, T., et al. 2013. *Measurements of methane emissions at natural gas production sites in the United States.* Proceedings of the National Academy of Sciences (PNAS) 500 Fifth Street, NW NAS 340 Washington, DC 20001, USA. October 29, 2013. 6 pgs. (http://www.pnas.org/content/early/2013/09/10/1304880110.full.pdf+html).

American Petroleum Institute (API) and America's Natural Gas Alliance (ANGA). 2012. *Characterizing Pivotal Sources of Methane Emissions from Natural Gas Production. Summary and Analysis of API and ANGA Survey Responses.* Final Report. September 21, 2012.

British Petroleum (BP). 2006. *Plunger Well Vent Reduction Project Producers Technology Transfer Workshop.* 2006. (http://www.epa.gov/gasstar/documents/desaulniers.pdf).

Drilling Information, Inc. (DI). 2012. *DI Desktop.* 2012 Production Information Database.

EC/R Incorporated. 2011. Memorandum to Bruce Moore, EPA/OAQPS/SPPD from Heather P. Brown, P.E., EC/R Incorporated. *Composition of Natural Gas for Use in the Oil and Natural Gas Sector Rulemaking.* July 28, 2011.

ICF International. 2011. *North American Midstream Infrastructure Through 2035 – A Secure Energy Future.* Prepared for the INGAA Foundation. 2011.

ICF International. 2014. *Economic Analysis of Methane Emission Reduction Opportunities in the U.S. Onshore Oil and Natural Gas Industries.* ICF International (Prepared for the Environmental Defense Fund). March 2014.

Lesair Environmental, Inc. 2008. *Oil & Gas Emissions Reduction Strategies Cost Analysis and Control Efficiency Determination.* June 2008.

Schlumberger. 1999. *Gas Lift Design and Technology.* Well Completions and Productivity Chevron Min Pass 313 Optimization Project. Pgs. I-12 - I-13.

U.S. Energy Information Administration (U.S. EIA). 2012a. Total Energy Annual Energy Review. Table 6.4 Natural Gas Gross Withdrawals and Natural Gas Well Productivity, Selected Years, 1960-2011. (http://www.eia.gov/total energy/data/annual/pdf/sec6_11.pdf).

U.S. Energy Information Administration (U.S. EIA). 2012b. Total Energy Annual Energy Review. Table 5.2 Crude Oil Production and Crude Oil Well Productivity, Selected Years, 1954-2011. (http://www.eia.gov/total energy/data/annual/pdf/sec5_9.pdf).

U.S. Energy Information Administration (U.S. EIA). 2013. Independent Statistics and Analysis. *Number of Producing Gas Wells.* (http://www.eia.gov/dnav/ng/ng_prod_wells_s1_a.htm).

U.S. Environmental Protection Agency. (U.S. EPA) 2006a. *Opportunities for Methane Emissions Reductions from Natural Gas Production.* Office of Air and Radiation: Natural Gas Star Program. Washington, DC. June 2006.

U.S. Environmental Protection Agency. (U.S. EPA) 2006b. *Installing Plunger Lift Systems In Gas Wells.* Office of Air and Radiation: Natural Gas Star Program. Washington, DC. 2006.

U.S. Environmental Protection Agency. (U.S. EPA) 2010. *Greenhouse Gas Emissions Reporting From the Petroleum and Natural Gas Industry: Background Technical Support Document.* Climate Change Division. Washington, DC. November 2010.

U.S. Environmental Protection Agency. (U.S. EPA) 2011. *Options for Removing Accumulated Fluid and Improving Flow in Gas Wells.* Office of Air and Radiation: Natural Gas Star Program. Washington, DC. 2011.

U.S. Environmental Protection Agency (U.S. EPA). 2012a. *Inventory of Greenhouse Gas Emissions and Sinks: 1990-2010.* Climate Change Division. Washington, DC. April 2012. (http://www.epa.gov/climatechange/ghgemissions/usinventoryreport/archive.html).

U.S. Environmental Protection Agency. (U.S. EPA) 2012b. *Technical Support Document: Federal Implementation Plan for Oil and Natural Gas Well Production Facilities. Fort Berthold Indian Reservation (Mandan, Hidatsa, and Arikara Nations), North Dakota. Attachment-FIP Emissions Control Cost Analysis from Operators.* 2012. EPA Region 8. EPA Docket No. EPA-R08-OAR-0479-0004.

U.S. Environmental Protection Agency. (U.S. EPA) 2013. *Petroleum and Natural Gas Systems: 2012 Data Summary. Greenhouse Gas Reporting Program.* October 2013. (http://www.epa.gov/ghgreporting/documents/pdf/2013/documents/SubpartW-2012-Data-Summary.pdf).

U.S. Environmental Protection Agency (U.S. EPA). 2014. *Inventory of Greenhouse Gas Emissions and Sinks: 1990-2011.* Climate Change Division. Washington, DC. April 2014. (http://www.epa.gov/climatechange/Downloads/ghgemissions/US-GHG-Inventory-2014-Chapter-3-Energy.pdf).